How Much
is an
ECOSYSTEM
Worth?
ASSESSING THE ECONOMIC VALUE OF CONSERVATION

SAVING THE LAST GREAT PLACES ON EARTH

THE WORLD BANK

ISBN 0-8213-6378-6
ISBN 13:   978-0-8213-6378-2
E-ISBN 0-8213-6379-4

Library of Congress cataloging-in-publication has been applied for.

Photos: Stefano Pagiola
Design: Jim Cantrell

CONTENTS

FOREWORD   v

ACKNOWLEDGMENTS   vii

ABBREVIATIONS   ix

1. INTRODUCTION   1

2. ECOSYSTEMS AND THE SERVICES THEY PROVIDE   5
  Ecosystem services   5
  Approaches to conservation   6

3. VALUING ECOSYSTEMS SERVICES   9
  Total economic value   9
  Valuation techniques   11

4. APPROACHES TO VALUATION   13
  Valuing the flow of current benefits   14
  Valuing the benefits and costs interventions that alter ecosystems   18
  Determining winners and losers   22
  Paying for conservation   24
  Summary   26

5. Conclusion   27

6. Further Reading   31

    The importance of ecosystems   31

    Valuation techniques: Theory   31

    Valuation techniques: Applications   32

    Sources for specific examples used in this paper   32

Boxes

Box 1: Making apples and oranges comparable   10

Box 2. Of diamonds and water   16

Box 3: "A serious underestimate of infinity"   17

Box 4: How much are pine kernels worth?   18

Box 5: Can benefits be transferred?   22

Box 6: Paying for watershed protection   26

Figures

Figure 1: Total Economic Value (TEV)   9

Figure 2: Flow of benefits from an ecosystem   14

Figure 3: Flow of benefits from forests in Mediterranean countries   15

Figure 4: Change in ecosystem benefits resulting from a conservation project   19

Figure 5: Cost-benefit analysis of a conservation project   20

Figure 6: Cost-benefit analysis of reforestation in coastal Croatia   21

Figure 7: Distribution of ecosystem benefits   23

Figure 8: Distribution of the costs and benefits of Madagascar's protected areas   24

Figure 9: Financing ecosystem conservation   25

Tables

Table 1: Main ecosystem types and their services   6

Table 2: Main economic valuation techniques   11

Table 3: Approaches to valuation   28

FOREWORD

The international community of nations has committed itself to achieve, by 2010, a significant reduction of the current rate of biodiversity loss at the global, regional, and national level. And yet despite growing awareness and major efforts in all countries, the latest evidence indicates that biodiversity continues to be lost at a terrifying pace, resulting in what some call the greatest mass extinction since dinosaurs roamed the planet, 65 million years ago.

There are many reasons for the gap between aspiration and reality. One of the most important is that economic policies and markets generally fail to value biodiversity or the conservation of ecosystems. With few exceptions, there is little financial reward for conserving biodiversity, nor much penalty for destroying it. Policy incentives to encourage nature conservation are emerging around the world, and yet this trend remains handicapped by a lack of understanding of the economic benefits of conserving natural ecosystems, or the costs of biodiversity loss.

A range of methods have been developed to value ecosystems and the services they provide, as well as the costs of conservation. The methods available are increasingly sensitive and robust, but they are often incorrectly used. One reason is poor understanding of the purposes of valuation and what questions it can, or cannot, answer. As a result, decision makers may get misleading guidance on the value of ecosystems and their conservation.

In this context, staff of the World Bank, IUCN—The World Conservation Union, and The Nature Conservancy have worked together to clarify the aims and uses of economic valuation, focusing on the types of questions that valuation can answer, and the type of valuation that is best suited to each purpose. This document is the result of that co-operation. It does not seek to provide a detailed "how to" manual on economic valuation. This report aims instead to provide guidance on how economic valuation can be used to address specific, policy-relevant questions about nature conservation.

We hope this document will stimulate discussions and assist decisions on using economic valuation for sustainable development. We welcome your comments.

Mr. Achim Steiner
Director General
IUCN—The World
Conservation Union

Mr. Steven J. McCormick
President and
Chief Executive Officer
The Nature Conservancy

Mr. Ian Johnson
Vice President
Environmentally and Socially
Sustainable Development
The World Bank

ACKNOWLEDGMENTS

This report was prepared by a joint team composed of Stefano Pagiola (World Bank), Konrad von Ritter (The Nature Conservancy), and Joshua Bishop (IUCN—The World Conservation Union). A more detailed technical paper is available separately. The team received valuable assistance and inputs from Kirk Hamilton, Gianni Ruta, and Patricia Silva (World Bank) and from Federico Castillo (University of California at Berkeley).

Several parts of this paper draw heavily from a draft manuscript on valuation of environmental impacts by Stefano Pagiola, Gayatri Acharya, and John Dixon. Permission to use this material is gratefully acknowledged.

The paper benefited from valuable comments provided by John Dixon (independent consultant), Edward Barbier (University of Wyoming), Randall Kramer (School of the Environment, Duke University), and Manrique Rojas (TNC).

ABBREVIATIONS

ANS	Adjusted Net Savings
CBD	Convention on Biological Diversity
CV	Contingent valuation
GDP	Gross domestic product
GNI	Gross national income
IRR	Internal rate of return
MA	Millennium Ecosystem Assessment
NPV	Net present value
NTFP	Non-timber forest product
PA	Protected area
PES	Payments for environmental services
TC	Travel cost
TEV	Total economic value
WTA	Willingness to accept
WTP	Willingness to pay

INTRODUCTION 1

The benefits provided by natural ecosystems are both widely recognized and poorly understood. What is increasingly clear, however, is that natural ecosystems are under enormous pressure around the world from the growing demands we place on them. Growth in human populations and prosperity translates into increased conversion of natural ecosystems to agricultural, industrial, or residential use, but also into increased demand for ecosystem inputs, such as fresh water, fiber, and soil fertility, as well as increased pressure on the capacity of natural ecosystems to assimilate our waste, including air and water pollution as well as solid waste. In short, we are asking more and more from natural ecosystems even as we reduce their capacity to meet our needs.

Stating that natural ecosystems and the services they provide are valuable immediately leads to the question: how valuable? This is an important question because other things are valuable as well. Maintaining ecosystems, whether through protected areas or through some other mechanism, requires expenditure of resources, and there are often many competing claims on these resources. Devoting more effort to conservation may mean having fewer resources to address other pressing needs, such as improving education, health, or infrastructure. Conserving ecosystems and the goods and services they provide may also involve foregoing certain uses of these ecosystems, and the benefits that would have been derived from those uses. Not converting a forest ecosystem to agriculture, for example, preserves certain valuable ecosystem services that forests may provide better than farmland, but it also prevents us from enjoying the benefits of agricultural production. To assess the consequences of different courses of action, it's not enough to know that ecosystems are valuable, we also need to know how valuable they are, and how that value is affected by different forms of management.

It has often been argued that a major reason for our failure to conserve natural ecosystems is that we do not realize how valuable they are. The farmers deciding whether to burn a hectare of forest to clear it for agriculture focus on the potential crop yields they may obtain, but pay little attention to the many ecological services that would go up in smoke. Likewise, national ministers of finance often base their budget decisions solely on the basis of indicators such as GDP, foreign exchange balances, and tax receipts, in which ecosystems services either do not appear or are not recognized as such—indeed, perversely, GDP often identifies activities that destroy ecosystems as 'benefits'. Not surprisingly, conservation budgets tend to get slighted.

Such concerns have led to an explosion of efforts to value natural ecosystems and the services they provide. The vast majority have focused on valuing only a sub-set of the benefits of particular ecosystems in specific locations (for example, the value of water filtration services provided by wetlands in Kampala, Uganda). Some more ambitious efforts have attempted to estimate the value of all services provided by broad categories of ecosystems (for example, the benefits of forests in Mediterranean countries), or even of all ecosystems on the planet.

Valuation studies have considerably increased our knowledge of the value of ecosystems, as well as of the strengths and limitations of different valuation methods. Another, less desirable outcome, however, has been growing confusion among decision-makers and non-economists about the validity and implications of ecosystem valuation. Unfortunately, environmental advocates in the media, government, business, and civil society have often seized on impressive but sometimes unsound valuation results and used them indiscriminately, and often inappropriately.

Valuation is not a single activity, and the seemingly simple question 'how valuable is an ecosystem?' can be interpreted in many different ways. It could be interpreted as asking about the value of the current flow of benefits provided by that ecosystem, for example, or about the value of future flows of benefits. It could also be asking about the value of conserving that ecosystem rather than converting it to some other use. These interpretations of the question are often treated as being synonymous, but they are in fact very different questions, and the answer to one will not be correct as an answer to the other.

Asking 'how valuable is an ecosystem?' also begs the question 'how valuable to whom?' The benefits provided by a given ecosystem often fall unequally across different groups. Ecosystem uses which seem highly valuable to one group may cause losses to another. Answering the question from the aggregate perspective of all groups (as is often the case in economic analysis), would thus give very different answers to answering it from the perspective of a particular group. Understanding the distribution of costs of benefits is also important when considering how to mobilize funds for conservation. Knowing that an ecosystem is valuable will not by itself ensure that it is conserved. Valuation can provide important insights into how conservation might be made financially sustainable—provided it is used the right way.

This paper seeks to clarify how valuation should be conducted to answer specific policy questions. In particular, it distinguishes four distinct approaches to valuing an ecosystem:

■ **Determining the value of the total flow of benefits from an ecosystem**. This question typically arises in a 'national accounts' context: How much is a given ecosystem contributing to economic activity? It is most often asked at the national level, but can also be asked at the global, regional, or local level.

■ **Determining the net benefits of interventions that alter ecosystem conditions.** This question typically arises in a project or policy context: Would the benefits of a given conservation investment, regulation, or incentive justify its costs? It differs fundamentally from the previous question in that it asks about *changes* in flows of costs and benefits, rather than the sum total value of flows.

■ **Examining how the costs and benefits of an ecosystem are distributed.** Different stakeholder groups often perceive very different costs and benefits from ecosystems. Understanding the magnitude and mix of net benefits received by particular groups is important for two reasons. From a practical perspective, groups that stand to 'lose' from conservation may seek to undermine it. Understanding which groups are motivated to conserve or destroy an ecosystem, and why, can help to design more effective conservation approaches. From an equity perspective, the impact of conservation on particular groups such as the poor, or indigenous peoples, is also often of significant concern in and of itself.

■ **Identifying potential financing sources for conservation.** Knowing that ecosystem services are valuable is of little use if it does not lead to real investments in conserving the natural ecosystems that provide them. Simply knowing that a protected area provides valuable watershed protection benefits, for example, does not pay the salaries of park rangers. Yet experience has shown that relying solely on government budget allocations or external donors for the necessary funding is risky. Valuation can help identify the main beneficiaries of conservation and the magnitude of the benefits they receive, and thus help design mechanisms to capture some of these benefits and contribute to financing of conservation.

These four approaches are closely related, but distinct. As will be shown in Chapter 4, they can be seen as looking at the same data from different perspectives. The specific answers to each of these questions can be very different, however, and the answer to one is often not meaningful when used as the answer to another.

The aim of this paper is not to provide detailed instructions on how to undertake valuation of ecosystem services, nor on how to use specific valuation techniques. There are many other sources that provide such instructions. Chapter 3 provides a summary of the main valuation techniques, their applicability to different problems, and their strengths and limitations. The section on Further Reading provides a wide variety of references to methodological references and to examples of their application. Rather, the objective of this paper is to clarify how valuation can and should be used to address important policy questions that often arise—and how such valuation differs from that which would be undertaken to address a different policy question. The interpretation of the results also differs.

Chapter 2 begins by providing a brief overview of the conservation problems we are addressing. These go beyond the narrow focus on protected areas that has often characterized the debate, and also include other conservation efforts. Although protected areas have been and will continue to be important tools for conservation, many valuable ecosystem services are provided by other land uses, including agriculture and industrial forestry. Chapter 3 provides a short summary of the main valuation techniques that have been developed to measure environmental benefits, their applicability to different problems, and their strengths and limitations. Chapter 4 then compares and contrasts the four approaches to valuation outlined above, showing how they differ and how they relate to each other. Chapter 5 concludes by discussing some of the limitations that valuation efforts face.

The view taken in this paper is that the purpose of valuation is to obtain reliable, objective information on the benefits and costs of conserving ecosystems so as to inform decisionmaking. In the context of evaluating a specific project or policy intervention, for example, it asks *whether* the resulting net benefits are sufficient to justify the costs of the intervention. The implication is that in some cases they may not. All too often, valuation is used merely as a tool to provide ammunition to support a predetermined position, with its results being discarded if they do not, in fact, support it.

We recognize that some people reject the assumptions and methods used to express environmental benefits in monetary terms. Although economic valuation methods are far from perfect, and are not the only way to assess ecosystem benefits, the view taken here is they are useful for illuminating trade-offs and guiding decisionmaking.

The focus of this paper is decidedly anthropocentric: the ecosystem benefits we consider are those that contribute to human well-being. This is not, of course, the only reason to be concerned about ecosystems. Many, drawing on a variety of ethical, philosophical, or cultural traditions, consider some or all ecosystems as having *intrinsic value*, whether or not they contribute to human well-being. There may be other reasons to conserve an ecosystem besides the economic benefits it provides. Understanding the economic costs and benefits of using ecosystems is thus only one of many inputs that enter into decisionmaking. The concern of this paper is that such understanding should be as accurate, meaningful, and useful as possible.

ECOSYSTEMS AND THE SERVICES THEY PROVIDE

Ecosystems, and biodiversity more generally, matter for many reasons. The reasons this paper focuses on are practical: ecosystems provide a wide variety of useful services that enhance human well-being. Without these services, we would be worse off in many ways. At the limit, we may not survive. But even degradation of ecosystem services falling well short of outright destruction would significantly affect our welfare.

Ecosystem services

The world's ecosystems provide a huge variety of goods and services. We are all familiar with the valuable commodities that natural ecosystems provide, such as edible plants and animals, medicinal products, and materials for construction or clothing. Many of us likewise value the aesthetic or cultural benefits provided by natural ecosystems, including beautiful views and recreational opportunities. What is less well known is the extent to which human economies depend upon natural ecosystems for a range of biological and chemical processes. These ecosystem 'services' are provided free-of-charge as a gift of nature. Examples of ecosystem services include the purification of air and water; regulation of rainwater run-off and drought; waste assimilation and detoxification; soil formation and maintenance; control of pests and disease; plant pollination; seed dispersal and nutrient cycling; maintaining biodiversity for agriculture, pharmaceutical research and development and other industrial processes; protection from harmful ultraviolet radiation; climate stabilization (for example, though carbon sequestration); and moderating extremes of temperature, wind, and waves.

Table 1 shows the world's major ecosystem types and the main services they provide, as described in the Millennium Ecosystem Assessment (MA). We follow here the MA's definition of ecosystems as dynamic complexes of plant, animal,

Table 1: Main ecosystem types and their services

Ecosystem service	Ecosystem									
	Cultivated	Dryland	Forest	Urban	Inland Water	Coastal	Marine	Polar	Mountain	Island
Freshwater			•		•	•		•	•	
Food	•	•	•	•	•	•	•	•	•	•
Timber, fuel, and fiber	•		•			•				
Novel products	•	•	•		•		•			
Biodiversity regulation	•	•	•	•	•	•	•	•	•	•
Nutrient cycling	•	•	•		•	•	•			
Air quality and climate	•	•	•	•	•	•	•	•	•	•
Human health		•	•	•	•	•				
Detoxification		•	•	•	•	•	•			
Natural hazard regulation			•		•	•			•	
Cultural and amenity	•	•	•	•	•	•	•	•	•	•

and microorganism communities and the non-living environment, interacting as functional units. It is important to note that this includes managed ecosystems such as agricultural land-scapes, and even urban areas. The MA classifies the services that ecosystem can provide into *provisioning services* such as food and water; *regulating services* such as flood and disease control; *cultural services* such as spiritual, recreational, and cultural benefits; and *supporting services*, such as nutrient cycling, that maintain the conditions for life on Earth. These categories illustrate the diverse ways in which ecosystems contribute to human well-being.

Despite the services they provide, natural ecosystems worldwide are under tremendous pressure. Forest ecosystems are being converted to other uses; wetlands are being drained; and coral reefs are being destroyed. Freshwater resources are increasingly modified through impoundment, redirection, extraction, land use changes that affect recharge and flow rates, and pollution. Agricultural soils and pasture lands are being degraded from over-use. Some of these pressures are intentional effects of human activities, others are un-intended.

Approaches to conservation

The standard approach to conservation has been the establishment of protected areas (PAs). This approach cordons off certain areas and restricts their use. There has been considerable debate about the effectiveness of PAs as instruments for protection. Recent research shows that PAs can be very effective in many cases. However, their effectiveness is limited by the fact that many PAs are too small and isolated to sustain the full range of ecosystem services. Moreover, due to weak capacity and limited resources many PAs are little more than 'paper parks'—protected in name only.

The limitations of PAs as a conservation strategy have led to increased attention being given to conservation efforts outside formally protected areas. Agricultural landscapes cover a large proportion of the world's surface, for example.

A variety of instruments have been developed to help improve conservation. As noted, the initial approach was a regulatory one, which sought to restrict land uses in particular areas. This approach includes the establishment of protected areas and rules that prohibit farming on sloping land or the use of pesticides in riparian areas. More recently, there have been increasing efforts to use market-based instruments to promote conservation. These approaches seek to change the behavior of land users by changing their incentives, thus encouraging them to adopt more environmentally benign land uses and discouraging them from adopting more harmful land uses. These approaches include efforts to develop markets for the products of environmentally-friendly land uses, such as shade-grown coffee; the purchase of easements or direct payments for conservation on private lands; and 'trading' systems designed to compensate for damage in one place by improvements elsewhere.

Whatever approach is used, conservation has both costs and benefits. The costs include both the direct costs of implementing conservation measures, and the opportunity costs of foregone uses. The benefits of conservation include preserving the services that ecosystems are providing—although it is important to note that not all conservation approaches conserve all services fully. The question thus immediately arises as to whether the benefits of a given conservation measure justify its costs.

VALUING ECOSYSTEMS SERVICES

Economic valuation offers a way to compare the diverse benefits and costs associated with ecosystems by attempting to measure them and expressing them in a common denominator—typically a monetary unit (see Box 1).

Total economic value

Economists typically classify ecosystem goods and services according to how they are used. The main framework used is the Total Economic Value (TEV) approach (Figure 1). The breakdown and terminology vary slightly from analyst to analyst, but generally include (i) direct use value; (ii) indirect use value; (iii) option value; and (iv) non-use value. The first three are generally referred to together as 'use value'.

- **Direct use values** refer to ecosystem goods and services that are used directly by human beings. They include the value of *consumptive uses* such as harvesting of food products, timber for fuel or construction, and medici-

Figure 1: Total Economic Value (TEV)

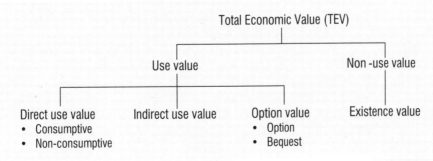

nal products and hunting of animals for consumption; and the value of *non-consumptive uses* such as the enjoyment of recreational and cultural activities that do not require harvesting of products. Direct use values are most often enjoyed by people visiting or residing in the ecosystem itself.

■ **Indirect use values** are derived from ecosystem services that provide benefits outside the ecosystem itself. Examples include natural water filtration which often benefits people far downstream, the storm protection function of mangrove forests which benefits costal properties and infrastructure, and carbon sequestration which benefits the entire global community by abating climate change.

■ **Option values** are derived from preserving the option to use in the future ecosystem goods and services that may not be used at present, either by oneself (*option value*) or by others/heirs (*bequest value*). Provisioning, regulating, and cultural services may all form part of option value to the extent that they are not used now but may be used in the future.

■ **Non-use values** refer to the enjoyment people may experience simply by knowing that a resource exists even if they never expect to use that resource directly themselves. This kind of value is usually known as *existence value* (or, sometimes, *passive use value*).

In general, direct use values are the easiest to value, since they usually involve observable quantities of products whose prices can usually also be observed in the market-place. Recreation is also relatively easy to value as the number of visits is directly observable. Assessing the benefit received by visitors is more difficult, but a large literature has developed to tackle this problem, mainly using surveys of tourists' actual travel costs or of their stated WTP to visit particular sites.

Measuring indirect use value is often considerably more difficult than measuring direct use values. For one thing, the 'quantities' of the service being provided—such as the amount of carbon stored in biomass or in the soil—are often hard to measure. While their contribution of ecosystem services to the production of marketed goods and services may be significant, it is often difficult to distinguish it from that of other, marketed inputs to production. Moreover, many of these services often do not enter markets at all,

---

Box 1: Making apples and oranges comparable

Valuation techniques typically express their results in monetary units. This is purely a matter of convenience, in that it uses units that are widely recognized, saves the effort of having to convert values already expressed in monetary terms into some other unit of account, and facilitates comparison with other activities that also contribute to well-being, such as spending on education or health. In particular, it expresses the impacts of changes in the services that ecosystems provide in terms of units that are readily understood by decisionmakers and the general public. When all impacts of ecosystem change are expressed in these terms, they can easily be introduced into frameworks such as cost-benefit analysis in order to assess and compare alternative courses of action. The use of monetary units to compare environmental values emphatically does *not* mean that only services which directly generate monetary benefits are taken into consideration in the valuation process. On the contrary, the essence of practically all work on the valuation of environmental and natural resources has been to find ways to measure benefits which do not enter markets and so have no directly observable monetary benefits.

Table 2: Main economic valuation techniques

Methodology	Approach	Applications	Data requirements	Limitations
Revealed preference methods				
Production function (also known as 'change in productivity')	Trace impact of change in ecosystem services on produced goods	Any impact that affects produced goods	Change in service; impact on production; net value of produced goods	Data on change in service and consequent impact on production often lacking
Cost of illness, human capital	Trace impact of change in ecosystem services on morbidity and mortality	Any impact that affects health (e.g. air or water pollution)	Change in service; impact on health (dose-response functions); cost of illness or value of life	Dose-response functions linking environmental conditions to health often lacking; under-estimates, as omits preferences for health; value of life cannot be estimated easily
Replacement cost (and variants, such as relocation cost)	Use cost of replacing the lost good or service	Any loss of goods or services	Extent of loss of goods or services, cost of replacing them	Tends to over-estimate actual value; should be used with extreme caution
Travel cost (TCM)	Derive demand curve from data on actual travel costs	Recreation	Survey to collect monetary and time costs of travel to destination, distance traveled	Limited to recreational benefits; hard to use when trips are to multiple destinations
Hedonic pricing	Extract effect of environmental factors on price of goods that include those factors	Air quality, scenic beauty, cultural benefits	Prices and characteristics of goods	Requires vast quantities of data; very sensitive to specification
Stated preference methods				
Contingent valuation (CV)	Ask respondents directly their WTP for a specified service	Any service	Survey that presents scenario and elicits WTP for specified service	Many potential sources of bias in responses; guidelines exist for reliable application
Choice modeling	Ask respondents to choose their preferred option from a set of alternatives with particular attributes	Any service	Survey of respondents	Similar to those of CV; analysis of the data generated is complex
Other methods				
Benefits transfer	Use results obtained in one context in a different context	Any for which suitable comparison studies are available	Valuation exercises at another, similar site	Can be very inaccurate, as many factors vary even when contexts seem 'similar'; should be used with extreme caution

so that their 'price' is also difficult to establish. The aesthetic benefits provided by a landscape, for example, are non-rival in consumption, meaning that they can be enjoyed by many people without necessarily detracting from the enjoyment of others.

Non-use value is the most difficult type of value to estimate, since in most cases it is not, by definition, reflected in people's behavior and is thus almost wholly unobservable (there are some exceptions, such as voluntary contributions that many people make to 'good causes', even when they expect little or no advantage to themselves). Surveys are used to estimate non-use or existence values, such as consumers' stated WTP for the conservation of endangered species or remote ecosystems which they themselves do not use or experience directly.

Valuation techniques

Many methods for measuring the economic value of ecosystem services are found in the resource and environmental economics literature (see the *further reading* section for references). Table 2 summarizes the main economic valuation techniques. Some are broadly applicable, some are applicable to specific issues, and some are tailored to particular data sources. A common feature of all methods of economic valuation of ecosystem services is that they are founded in the theoretical axioms and principles of welfare economics. Most valuation methods measure the demand for a good or service in monetary terms, that is, consumers' willingness to pay (WTP) for a particular benefit, or their willingness to accept (WTA) compensation for its loss.

Some techniques, generally known as 're-vealed preference' techniques, are based on ob-served behavior. These include methods that de-duce values indirectly from people's behavior in surrogate markets, which are hypothesized to be related to the ecosystem service of interest. Other techniques are based on hypothetical rather than actual behavior, where people's responses to questions describing hypothetical situations are used to infer their preferences. These are generally known as 'stated preference' techniques. In general, measures based on observed behavior are preferred to measures based on hypothetical behavior, and more direct measures are preferred to indirect measures. However, the choice of valuation technique in any given instance will be dictated by the characteristics of the case and by data availability.

It is important to use these valuation techniques properly. They provide powerful tools to assess the value of particular ecosystem benefits, but if they are mis-applied their results will be of little use. This paper does not provide detailed guidance on using these techniques; many other sources do so (see *Further Reading* section). Rather, the purpose of this paper is to help decision-makers frame the valuation question properly to ensure that the numbers these techniques provide are relevant and useful for addressing specific policy issues.

APPROACHES TO VALUATION 4

Consider a minister of finance. The minister of environment has been asking her for an increase in the budget for protected areas—partly to improve conservation in areas that are currently just 'paper parks' and partly to expand the protected area system so as to include many unique ecosystems which are currently omitted. Small farmers are encroaching into protected areas, he warns, burning down the forests to clear land for agriculture. At the same time, the minister of education has been complaining about insufficient resources to achieve the goal of universal primary education, and keeps asking "why protect butterflies when our children cannot read?" A delegation of agro-industrialists wants to convert large swathes of land to soybean production for export, and is requesting financing for a road to improve their access to ports. The national electric company also needs financing, to dredge a reservoir that has been filled with sediment, reducing hydroelec-

tric power generating capacity. How does one proceed in this welter of conflicting demands?

Experts from each group can easily churn out reams of evidence to back their position. The agro-industrialists speak of production increases, jobs, and exports. The electric company speaks of megawatt hours. The minister of environment tends to speak of species and endemism, but also brings up erosion, flood risk, pollination, and tourism. The small farmers are not at this table, but they are known to need food and income for their subsistence, and if they cannot get land here they'll look for it elsewhere, or join the throngs already crowding urban slums.

How does one make sense of these competing needs? How can the minister prioritize if it entails comparing apples to oranges? How can the minister of finance see her way through to a rational decision about where to invest scarce public funds? Economic valuation offers part of the answer, by attempting to measure all the diverse benefits and costs associated with ecosystems and expressing them in a common denominator.

For the numbers to be useful, however, they must not only be comparable, they must also measure the right thing and measure them in the right way.

Valuing the total flow of benefits

One of the concerns of the minister of finance in our hypothetical country is to have an accurate assessment of the state of the national economy at any point in time. She and her President, as well as international agencies and foreign banks, are constantly reviewing their policies and investment decisions in light of changing economic conditions. Is the economy growing too rapidly or too slowly? Is the burden of public debt sustainable? What are the prospects for growth in different sectors of the economy? These and other macroeconomic questions can only be answered with a reliable set of economic indicators.

Unfortunately, the information available to the minister of finance is seriously incomplete. She knows very well that her country relies heavily on agriculture, tourism, fishing and forestry for both job creation and export income. She is likewise constantly reminded, by her colleague from the environment ministry, that these activities depend on natural ecosystems to provide fertile soils for crop production, lush vegetation and wildlife to attract foreign tourists, clean water and healthy mangrove swamps to sustain inland and off-shore fisheries, and healthy forests for the production of timber, fuelwood, and other commodities. And yet, despite the importance of ecosystem services to the economy, their contribution is hard to discern in the available statistics.

Some ecosystem benefits appear in national accounts (such as many extractive uses), but many either do not appear at all (most non-use values and many indirect uses) or are hidden in the benefits ascribed to other parts of the economy (including many indirect use values). As a

Figure 2: Flow of benefits from an ecosystem

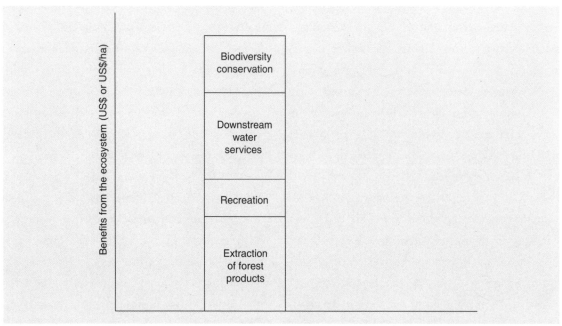

result, the benefits provided by natural ecosystems appear to be much smaller than they really are. Estimating the economic value of the benefits provided by natural ecosystems can provide a much more accurate sense of how important those ecosystems are to the economy. If the value of the various benefits could be estimated, the result would look something like Figure 2, which could be presented either in terms of total benefits of that ecosystem, or in terms of per hectare benefits. An estimate in total terms would be better suited to a comparison to GDP numbers, while an estimate expressed in per hectare terms would be better suited to comparison to alternative land uses.

Such estimates can be useful in demonstrating that seemingly 'worthless' land uses may in fact be quite important to the economy. They can clarify the relative importance of ecosystem services to total economic output, and thus guide overall investment strategy, although more detailed analysis is usually required to assess specific interventions, as discussed below.

Figure 3 illustrates the results of one such effort, which sought to value the current flow of benefits provided by countries in the Mediterranean basin. This study estimated the average value of these flows to about US$150/ha a year. This is likely to be an underestimate, however, as many non-market benefits could not be estimated in many cases. The gap between the estimated TEV in European countries and that in North African and Middle Eastern countries is probably smaller than it appears here, as data constraints were particularly severe in the latter countries. Note that some flows are actually dis-

Figure 3: Flow of benefits from forests in Mediterranean countries

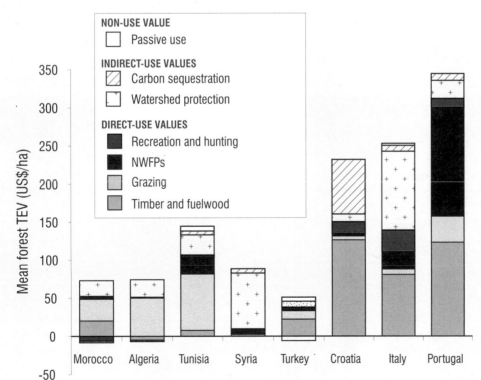

benefits: erosion from poorly managed portions of the forest, for example, is imposing costs on downstream populations.

Such studies provide two important insights. First, they show how important ecosystems are to society. On a per capita basis, forests in Mediterranean countries (Figure 3) provide at least US$50 annually—about US$70 per capita in European countries, but less than US$10 in North African and Middle Eastern countries, with their much smaller forest areas. On average, forest benefits in the region amount to about 1 percent of GDP. As noted above, these are likely to be lower-bound estimates. Second, the composition of total flows tells us how likely it is that ecosystems are being used sustainably. Direct use values are generally enjoyed by local people, and so they generally have incentives to manage them sustainably—although this is not necessarily true if tenure is insecure, as is often the case in forest areas. But even with secure land use rights, local people have little incentive to protect indirect use values. In the case of Mediterranean forests, indirect use values such as watershed protection contribute about 35 percent of the total estimat-

ed value, and even this share is probably underestimated, as it is harder to measure indirect uses than direct uses. Thus a very large proportion of the services that these ecosystems provide are likely to benefit people other than those who directly manage them—a situation quite likely to result in sub-optimal resource use.

One should be careful in interpreting estimates of the total flows of benefits from an ecosystem. If the ecosystem is not being managed sustainably, these flows may well decline in the future. High rates of extraction of products such as timber, for example, may not be sustainable if extraction exceeds the natural rate of growth. High apparent flows of current benefits may thus come at the expense of future flows. (Conversely, if extraction is less than natural growth, the stock of the resource would grow over time; in this case the current flow of benefits would tend to understate potential future benefits.)

Interpretation is also difficult when examining the value of ecosystems on a large scale. The thought experiment that underlies this approach to valuation is 'how much worse off would we be if we did not have the ecosystem,

Box 2. Of diamonds and water

There is a well-known paradox in economics called the 'diamonds and water paradox': Water, despite its importance for survival itself, is generally very cheap, while diamonds, despite their relative unimportance except as an adornment, tend to be very expensive. The reason for this paradox lies in the relative abundance of water and diamonds. Water is generally plentiful, and so an additional unit tends to be cheap. Diamonds, on the other hand, are scarce, and so command a high price.

How is this relevant to the valuation of ecosystem services? Most valuation studies of services such as water supply have been undertaken in contexts where these services are relatively abundant. Even in cases where water is considered scarce (usually defined as availability of less than 1,000m^3/capita/year) it is not usually so scarce as to endanger life itself. Using these results to estimate the value of services provided by small-scale ecosystems is appropriate. But as we start considering the value of all services provided by ecosystems on a large scale, the premises of the paradox no longer hold. If we had no water at all, it would be extraordinarily valuable. So if we consider the value of all water provided by a large ecosystem, or all freshwater on the planet, the marginal price for an extra unit of water is no longer a reliable guide.

and all the services it provides, at all?' This question is reasonably well defined at small scales (a particular forest, say, in a specific country): the total physical flow of goods such as timber is small enough that its presence or absence is unlikely to affect prices. As the scale of the analysis increases (to cover, for example, all tropical forests, or an entire country), this assumption is less and less likely to hold, and observed prices are less and less likely to be applicable (see Box 2). This is one of the reasons that efforts to value all the world's ecosystems can produce non-sensical results (see Box 3).

It should also be noted that the results of this kind of analysis apply to the ecosystem as it is currently being managed. Different management practices would result in a different flow of benefits.

Efforts to value 'natural capital' are a variation of this approach. Rather than looking at the flow of benefits from an ecosystem in a single year, the natural capital approach considers the present value of all current and future benefits that the ecosystem will generate. Estimating this value requires projecting how the flow of services, and their value, would evolve over time.

Where multi-year estimates of ecosystem service values are available, they can support the construction of more reliable indicators of national economic performance, such as Adjusted Net Savings (ANS). ANS measures the change in total wealth in a given period; it is calculated by adjusting the traditional measure of net national savings to account for activities which enhance wealth, such as education expenditure (an investment in human capital), as well as activities that

Box 3: "A serious underestimate of infinity"

A landmark paper published in *Nature* in 1997 attempted to calculate the total value of all ecosystems on earth. By using a range of estimates of the value of individual ecosystems and scaling them up according to the total area covered by each such ecosystem globally, the authors arrived at an estimate of the total value of all ecosystem services ranging from US$16-54 trillion a year, with a central estimate of US$33 trillion (in 1997 prices).

This paper has had a significant impact and its results have been widely quoted by scientists and environmentalists. However, most economists consider it profoundly flawed, both conceptually and methodologically.

- The study generates its global estimates by using the results of valuation studies undertaken in specific locations and extrapolating them to other areas. As discussed in Box 5 and illustrated in Figure 6, such 'benefits transfers' are often unreliable due to wide variation in ecosystem values across different sites.

- The study uses estimates of average value based on marginal changes in ecosystem services to calculate the aggregate value of entire ecosystems. However, this approach fails to account for the variation in unit values as the scale of analysis changes. As discussed in Box 2, the value ascribed to a resource depends on whether the change in availability we are contemplating is large or small.

- The study results exceed the sum total of global economic income recorded in 1997. They cannot be interpreted as an estimate of society's willingness to pay, as one cannot plausibly pay more than the total value of all income. The results also cannot be interpreted as society's willingness to accept compensation for loss of all the world's ecosystem services. Without these services, we would all be dead, so there is no finite compensation we would accept for the loss of all ecosystem services. In this sense, the results are "a serious underestimate of infinity," in the words of Michael Toman.

- The study suggests that its results should guide policy decisions. But, as discussed throughout this paper, information on total benefit flows, even if accurate, cannot provide guidance on specific conservation decisions, which are about making incremental changes in those flows.

reduce wealth, such as depletion of mineral and energy reserves, forest depletion, and damages from pollution. The adjustment is often considerable. In the case of Ghana, for example, traditional measures show a saving rate of over 15 percent of gross national income, but adjusted net savings of only about 6 percent. Depletion of forests accounts for about 3 percentage points of this adjustment.

Valuing the benefits and costs interventions that alter ecosystems

Our hypothetical finance minister has commissioned several rigorous studies of the value of various ecosystems in her country. She has been told that one particular area of natural forest generates significant economic benefits in the form of wood and non-wood products extracted by local communities, as well as valuable services such as recreation and protection of water supplies to the capital city—indeed, these benefits amount to a non-negligible portion of GNP for the region in which the forest is located. Conser-

vation of the forest would seem a sensible course of action and the minister of environment has submitted a comprehensive proposal to enhance the watershed, wildlife, and recreational benefits it provides, while also reducing unsustainable poaching and harvesting of forest products. The proposal is impressive but it is also expensive. Moreover, this is the same area that the agroindustrialists are targeting for soybean production. Protecting the forest means foregoing this project. What should the finance minister do?

Estimates of the total annual flow of benefits from an ecosystem have frequently been used to justify spending to address threats or to improve its condition. But using such value estimates in this way would be a mistake. To examine the consequences of ecosystem degradation, or to assess the benefits of a conservation intervention, it is not enough to know the total flow of benefits. Rather, what is needed is information on how that flow of benefits would *change*. It is rare for all ecosystem services to be lost entirely, even if a natural habitat is severely degraded: a forested watershed that is logged and converted to agriculture, for example, will still provide a mix of environmental services, even though both the mix and the magnitude of specific services will have changed. It would be a mistake, therefore, to credit a conservation project which prevents such degradation with the *total* value of the flow of benefits provided by the ecosystem at risk. Rather, what is needed is an assessment of the incremental change in the value of services provided by the ecosystem resulting from a well-defined change in how it is managed. This is illustrated in Figure 4. A critical point illustrated in the figure is that this analysis should not compare ecosystem benefits before and after conservation measures are implemented, as many other factors may also have changed in the intervening period. Rather, it should compare ecosystem benefits with and without the conserva-

tion measures: that is, it must compare what would happen if conservation measures were implemented to what would have happened if they were not.

Such an assessment can be undertaken either by explicitly estimating the change in value arising from a change in management, or by separately estimating the value of ecosystem services under the current and alternative management regimes, and then comparing them. If the loss of a particular ecosystem service is irreversible, then the loss of the option value of that service should also be included in the analysis.

The minister's question, then, is whether the total economic value of the services provided by an ecosystem managed in one way (with conservation) is more or less than the total value generated by the ecosystem if it were managed

Figure 4: Change in ecosystem benefits resulting from a conservation project

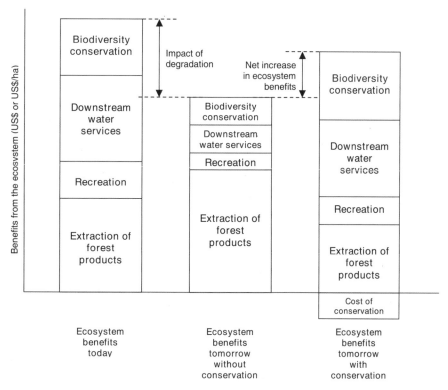

in another way (without conservation). It is quite likely that a change in management will increase the value of some services and decrease the value of others; what matters is the net difference between the total value of all services, as shown in Figure 4. This difference is the value of conservation and is not necessarily positive. In other words, there may be cases in which the value of the additional services obtained by converting an ecosystem to an alternative use exceeds the value of the services obtained under conservation. The change in value must then be compared to its cost in order to determine whether it is worth undertaking, from an economic perspective.

Figure 5 illustrates this same approach in a different way. Here the values of the various services that would be obtained with and without conservation are compared directly. Some servic-

es are increased thanks to conservation, while others (in this case, extraction of forest products) are reduced. The third column shows the net changes in each service, along with the cost of conservation. This presentation illustrates the fact that the cost of conservation actually has two components: the direct, out-of-pocket costs of implementing the conservation measures themselves, and the opportunity cost of the foregone benefits from the services whose use is restricted. These two costs should then be compared to the gross increase in ecosystem benefits that would result from implementing the conservation measures. It is a very common mistake to consider only the out-of-pocket costs of conservation, ignoring the opportunity costs.

Figure 6 illustrates the results of one such analysis, for a proposed reforestation project in

Figure 5: Cost-benefit analysis of a conservation project

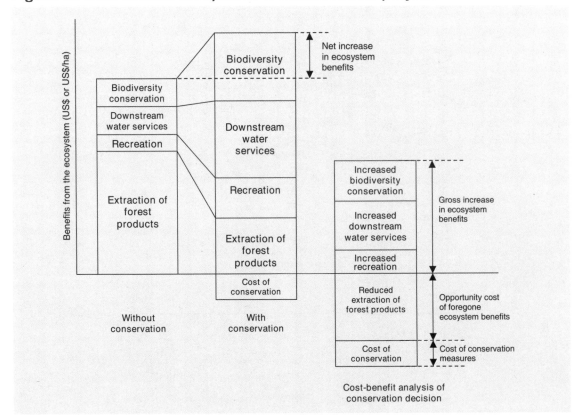

Figure 6: Cost-benefit analysis of reforestation in coastal Croatia

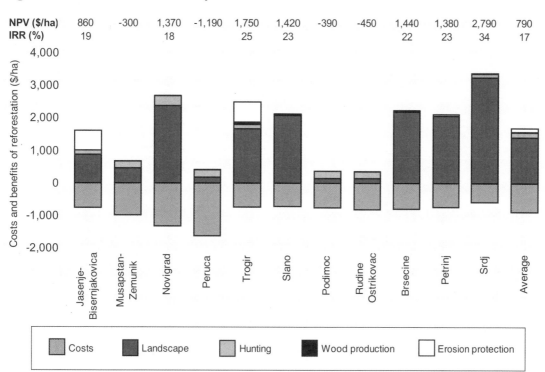

| NPV ($/ha) | 860 | -300 | 1,370 | -1,190 | 1,750 | 1,420 | -390 | -450 | 1,440 | 1,380 | 2,790 | 790 |
| IRR (%) | 19 | | 18 | | 25 | 23 | | | 22 | 23 | 34 | 17 |

coastal Croatia. The results are presented in the same format as the last column in Figure 5: as changes in each of the services resulting from implementing the proposed conservation project. The results show that both costs and benefits vary by site, depending on local characteristics. Sites on steeper slopes, for example, are costlier to reforest. Thus even within the same county, benefits can vary by several orders of magnitude: the average benefit of US$790/ha (discounted at 10 percent) masks substantial variation in the net benefits of reforestation.

Estimating changes in ecosystem benefits and costs is sometimes easier than estimating the value of the total flow of benefits of an ecosystem, because the analysis can focus on only those benefits and costs which are affected by the proposed conservation action. Often, many of the services provided by an ecosystem will be rela-

tively unaffected by a given intervention. In the case of Croatia, for example, many of the proposed reforestation sites had no downstream facilities at risk from erosion. At those sites, therefore, erosion reduction benefits could be ignored, as they were negligible. On the other hand, estimating the changes in physical flows of benefits as a result of an intervention is sometimes more difficult than estimating total flows.

Determining winners and losers

The discussion thus far has focused on aggregate benefits and costs. If the increase in aggregate benefits exceeds the increase in aggregate costs, then conservation would be interpreted as being worthwhile from society's perspective. But our Finance Minister is also concerned about who will receive the benefits and who will bear the costs. Consideration of aggregate benefits and costs masks the fact that those benefits and costs can be distributed very un-evenly across groups. This is illustrated in Figure 7, in which benefits are broken down into three groups: those received by local users, those received by the rest of the country, and those received by the global community. A similar analysis could be conducted showing how benefits would change as a result of a conservation intervention.

The un-even distribution of costs and benefits has both practical and ethical consequences. In practical terms, it is important to understand the costs and benefits received by local users, as they often have a very strong influence on how the ecosystem is managed. If local users stand to gain more from a particular land use, they may well convert the ecosystem to that land use no matter how large the benefits of conservation are to others. Likewise, if local users stand to benefit more from current conditions than from a proposed intervention, they are likely to oppose that intervention. Understanding who gains and—in particular—who loses from ecosystem

conservation thus provides important insights into the incentives that different groups have to manage an ecosystem in a particular way. By comparing the net benefits that groups receive from an ecosystem managed in one way (without conservation, say) to the net benefits they would receive if it were managed in another way (with conservation), this approach can also help predict which groups are likely to support a change in management, and which groups are likely to oppose it. This approach can thus provide useful information in the design of appropriate responses.

More fundamentally, distributional analysis is important to ensure that management interventions do not harm vulnerable people, and to design interventions that help reduce poverty and social exclusion. Tracking the flow of costs and benefits to different stakeholder groups allows us to understand how conservation actions affect the poor and other groups of interest, such as indigenous peoples. In the past, conservation efforts such as the creation of protected areas have

Box 5: Can benefits be transferred?

'Benefits transfer' refers to the use of valuation estimates obtained (by whatever method) in one context to estimate values in a different context. For example, an estimate of the benefit obtained by tourists viewing wildlife in one park might be used to estimate the benefit obtained from viewing wildlife in a different park. Alternatively, the relationship used to estimate the benefits in one case might be applied in another, in conjunction with some data from the site of interest ('benefit function transfer'). For example, a relationship that estimates tourism benefits in one park, based in part on tourist attributes such as income or national origin, could be applied in another park, using data on income and national origin of the latter park's visitors.

Benefits transfer is a seductive approach, as it is cheap and fast. It has been the subject of considerable controversy in the economics literature, however, as it has often been used inappropriately. Figure 6 illustrates how dangerous it can be: even within a narrowly-defined environment (forests in coastal Croatia, an area of about 5,000 km^2), the benefits of ecosystem services can differ by an order of magnitude. A consensus seems to be emerging that benefit transfer can provide valid and reliable estimates under certain conditions. These include the requirement that the commodity or service being valued should be very similar at the site where the original estimates were made and the site where they are applied; and that the populations affected should also have very similar characteristics. Of course, the original estimates must themselves be reliable for any attempt at transfer to be meaningful.

Figure 7: Distribution of ecosystem benefits

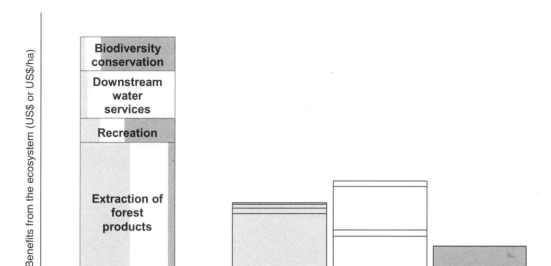

often had a negative impact on many local communities, for example by reducing their access to resources upon which they depend for their livelihood. Such impacts are of greatest concern where the affected population is most deprived: even if the economic cost is small compared to the overall benefits, it could be very significant for poor households. Identifying and estimating the value of such impacts can allow for conservation strategies to be modified to avoid them, or for appropriate compensation mechanisms to be designed.

Valuing the benefits and costs obtained by individual groups obviously requires identifying which specific services they use. An initial breakdown that is useful in many cases is between local groups, the rest of the nation, and the rest of humanity. In many cases, however, it is necessary to subdivide groups more finely: rubber tappers and loggers may both be local, for example,

but they derive very different benefits from a forest and thus have different interests.

Figure 8 illustrates the results of one such analysis of the costs and benefits of Madagascar's protected area system. The first column summarizes the overall benefits of the protected area system to Madagascar. Despite the high management costs and the foregone income from use of that land, the protected area system is estimated to provide net benefits to the country, thanks to the valuable watershed protection services these areas provide, their tourism benefits, and payments received from the global community for protection of the country's unique biodiversity. But as the breakdown in the right side of the figure shows, these benefits are very unevenly distributed. Local communities bear the brunt of the costs, as they are barred from using protected areas either for agriculture or for the collection of fuelwood and other non-timber forest prod-

Figure 8: Distribution of the costs and benefits of Madagascar's protected areas

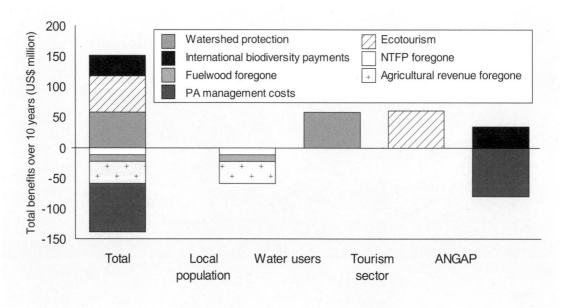

ucts (NTFPs). Downstream water users such as irrigated farmers benefit substantially, as do tourism operators. The protected area management agency, ANGAP, bears the management costs but receives external support (and a part of the tourism benefits). These results indicated the need for support to protected areas to include appropriate compensation mechanisms for local communities.

Paying for conservation

Even if our minister of finance was persuaded by the economic valuation study that the country as a whole would benefit from an intervention or policy to maintain the current level of ecosystem goods and services, she will still have to answer the question of how to finance those interventions. In particular, effective conservation usually requires a long-term commitment of resources. Almost invariably, the resources available for conservation are grossly inadequate to the task. Thus, even if conservation could, in prin-

ciple, generate large economic benefits, it often does not happen. Or, more commonly, it happens for some time, thanks to funding from a donor, and then collapses once the project and its funding come to an end. Identifying potential financing sources for conservation is thus another important use for valuation.

Economic valuation can help make conservation financially sustainable in two ways. First, by demonstrating the benefits that ecosystems generate, and the increased benefits (or avoided losses) that conserving these ecosystems can bring to stakeholders, valuation can help convince our finance minister and other decision-makers to allocate more resources to conservation. The study of Madagascar illustrated in Figure 8 played an important role in this regard, for example. But it would be overly optimistic to expect all problems to be resolved this way. Second, valuation can provide invaluable support to these efforts by identifying and quantifying the major benefits provided by a given ecosys-

Figure 9: Financing ecosystem conservation

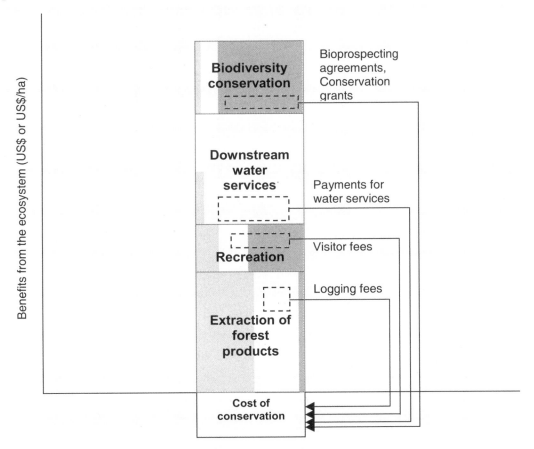

tem, and by identifying who the beneficiaries are. Based on this information, a variety of approaches might be used to secure additional funding for ecosystem conservation, as illustrated in Figure 9. When recreational use is important, for example, there is often a potential to use entrance fees. It is important to know to what degree recreational use is undertaken by foreigners as opposed to nationals, as this may affect viable fee levels. Such mechanisms are already widely used, although fee levels are typically set far below their potential.

Some extractive uses are also susceptible to being tapped for increased funding. It may be difficult to charge local users for collecting fuelwood and NTFPs, but relatively easy to increase

royalties paid by loggers. Likewise, when valuation shows that indirect uses such as watershed protection provide important benefits, then payments for environmental services (PES) are a promising approach. In a PES program, downstream water users pay fees which are used to finance payments to land users in upper watersheds who undertake appropriate land uses. Several cities and towns have implemented such programs (Box 6). Other PES programs focus on carbon sequestration services. Biodiversity benefits are often the hardest to capture, but even here there has been considerable experimentation with a variety of approaches. Note that it is often unrealistic to expect to capture the entire benefit from various user groups. As illustrated in Fig-

Box 6: Paying for watershed protection

Recent years have seen an increasing use of mechanisms based on the principles of payments for environmental services (PES), particularly in Latin America. Costa Rica and Mexico have created nationwide PES programs. The vast majority of PES initiatives, however, have been for smaller-scale initiatives at the scale of individual watersheds. Irrigation water user groups, municipal water supply systems, and hydroelectric power producers in several countries participate in such programs. The cities and towns that use PES to protect their water supplies cover a wide spectrum, from Quito, Ecuador, with 1.2 million people, to Yamabal, El Salvador, with only 3,800 people.

Valuation is a critical step in the development of PES programs. The payments must obviously exceed the additional benefit to land users of the alternative land use (or they would not change their behavior) and less than the value of the benefit to downstream populations (or they would not be willing to pay for it). Without valuation, it may be difficult to set an appropriate payment level, or even to determine whether the program is worth implementing at all.

ure 9, typically only a portion of their benefit can be captured.

Summary

The four approaches described here are closely linked and build on each other. They represent four different ways to look at similar data regarding the value of an ecosystem: its total value or contribution to society, the change in this value if a conservation action is undertaken, how this change affects different stakeholders—that is, who are the beneficiaries and who are the losers—and how beneficiaries could be made to pay for the services they receive to ensure that the ecosystem is conserved and its services are sustained. Each of these approaches to valuation uses similar data. They use that data in very different ways, however, sometimes looking at all of it, sometimes at a subset, sometimes looking at a snapshot, and sometimes looking at changes over time. Each approach has its uses and its limitations. Understanding under what conditions one approach should be used rather than another is critical: the answer obtained under one approach, no matter how well conducted, is generally meaningless when applied to problems that are better treated using another approach. In particular, using estimates of total flows to justify specific conservation decisions—although commonly done—is almost always wrong, though common. Properly used, however, valuation can provide invaluable insights into conservation issues.

CONCLUSION 5

Only a few decades ago, many standard economics textbooks considered environmental costs and benefits to be 'unquantifiable' and advised against trying to estimate them. Since then, tremendous progress has been made in developing a range of techniques for valuing environmental costs and benefits. Today our toolkit is well-stocked with increasingly sensitive and generally reliable valuation method. There is a growing body of literature that applies these techniques to a wide range of environmental issues. There is no longer any excuse for considering environmental costs and benefits as unquantifiable.

Economic valuation can provide useful information—when it is donecorrectly. Valuation of ecosystem goods and services can seem deceptively simple—a matter of multiplying a price by a quantity. In practice, however, valuation is often complex and normally requires specialist training and experience to ensure credible results.

But even expert economists will produce information that is of little use if the questions they are asked are badly framed. As this paper has sought to clarify, different policy contexts require different approaches. In particular, estimates of the total annual flow of benefits from an ecosystem, while often very impressive, are a poor guide to policy and investment decisions. Much more useful in most cases are estimates of the changes in benefit flows that will result from changes in ecosystem management. Table 3 summarizes the main characteristics of each approach.

Economic analysis is not and should not be the only input into conservation decisions. People can and do decide to conserve things based on a range of other criteria, such as for ethical, cultural, and historical reasons. Even then, valuation can provide relevant information—for example, by highlighting the economic consequences of alternative courses of action. Thus economic

Table 3: Approaches to valuation

Approach	Why do we do it?	How do we do it?
Determining the total value of the current flow of benefits from an ecosystem	To understand the contribution that ecosystems make to society	Identify all mutually-compatible services provided; measure the quantity of each service provided; multiply by the value of each service
Determining the net benefits of an intervention that alters ecosystem conditions	To assess whether the intervention is economically worthwhile	Measure how the quantity of each service would *change* as a result of the intervention, as compared to their quantity without the intervention; multiply by the marginal value of each service
Examining how the costs and benefits of an ecosystem (or an intervention) are distributed	To identify winners and losers, for ethical and practical reasons	Identify relevant stakeholder groups; determine which specific services they use and the value of those services to that group (or changes in values resulting from an intervention)
Identifying potential financing sources for conservation	To help make ecosystem conservation financially self-sustaining	Identify groups that receive large benefit flows, from which funds could be extracted using various mechanisms

valuation, used correctly, will lead to more informed choices even when economic considerations are not the primary criterion for decision-making.

It is rarely feasible or desirable to estimate every environmental benefit or cost. Even where valuation provides only partial results, however, it can help to structure how we think about conservation, identify critical information gaps and clarify the relation between ecosystem processes and human well-being. Indeed, an important benefit of attempting to undertake economic analysis is that it forces us to grapple with our limited understanding of ecosystem processes and the way they affect human well-being. All too often, public debate and policy on conservation is based on vague statements about ecosystems benefits, which implicitly assign a value of either zero or infinity to natural ecosystems. Zero is clearly wrong, but infinity is equally unhelpful as it prevents us from setting priorities. The

types of analyses discussed in this paper force us to be explicit about our assumptions: what specific services does an ecosystem provide? Who receives those services? How important are they? How would each of these services change if the ecosystem were managed differently? How big would the change be? How rapid? How long-lasting? Would it be reversible? What substitutes exist, if any? Simply stating the questions involved in an economic valuation can help to identify what we know and what we don't know about the role that ecosystems play in our well-being.

While valuation can shed useful light on many issues, there are several questions that economic valuation techniques handle poorly. Most of the direct and indirect use values of ecosystems can be measured quite accurately and reliably—the main constraint is often the availability of relevant physical data (that is, information on the quantity of service provided, or on the change in the quantity of service provided) rather than economic data (on the value of an extra unit of the service). Estimating option values and existence values involves greater uncertainties. Valuation of changes in human mortality is also problematic, as many people (including many economists!) find the notion of assigning a monetary value to human life unacceptable.

Economic valuation also tends to handle very large-scale and long-term problems rather poorly. Existing economic valuation techniques can provide reliable answers to questions involving relatively small-scale changes in resource use or availability, but become less robust as the scale of the analysis and the magnitude of environmental change increases. Similarly, economic valuation tends to deal poorly with very long time horizons. Uncertainty about future benefit flows becomes increasingly important, and the role of discounting increasingly determinant. Alternative non-economic approaches such as setting Safe Minimum Standard (SMS) may be more suitable in such cases, particularly when changes are thought to be irreversible.

Economic valuation has both strengths and limitations as a tool for decisionmaking. It is clear, however, that decisions about environmental management are not getting easier, and that information about costs and benefits is increasingly essential to ensure efficient, equitable, and sustainable outcomes. Valuation can play an important role in providing such information, provided it is used correctly.

FURTHER READING

This paper is based on a longer and more comprehensive background paper. A full list of references and citations for sources used in preparing this paper can be found there.

Pagiola, S., K. von Ritter, and J.T. Bishop. 2004. "Assessing the Economic Value of Ecosystem Conservation." Environment Department Paper No.101. Washington: World Bank.

A companion CD-ROM also provides a variety of examples of applications of economic valuation techniques.

Valuing Ecosystem Benefits: Readings and Case studies on the economic value of exosystem conservation.

This section provides a selected list of further readings. It is not intended to be comprehensive, but to provide an initial set of readings for readers interested in pursuing specific subjects in more depth.

The importance of ecosystems

G. Daily (ed.). 1997. *Nature's Services: Societal Dependence on Natural Ecosystems*. Washington: Island Press.

Millennium Ecosystem Assessment, 2003. *Ecosystems and Human Well-being: A Framework for Assessment*. Washington: Island Press.

World Resources Institute. 2000. *World Resources 2000-2001: People and Ecosystems: The Fraying Web of Life*. Washington: World Resources Institute.

Valuation techniques: Theory

Underlying economic theory

Braden, J.B. and C.D. Kolstad (eds.). 1991. *Measuring the Demand for Environmental Quality*. Contributions to Economic Analysis No.198. Amsterdam: North Holland.

History of economic valuation

Hanemann, W.M. 1992. "Preface." In S. Navrud (ed.), *Pricing the European Environment*. Oslo: Scandinavian University Press.

Pearce, D. 2002. "An intellectual history of environmental economics." *Annual Review of Energy and the Environment*, **27**, pp.57-81.

General overviews

Dixon, J.A., L.F. Scura, R.A. Carpenter, and P.B. Sherman. 1994. *Economic Analysis of Environmental Impacts.* London: Earthscan.

Freeman, A.M. 1993. *The Measurement of Environmental and Resource Values: Theory and Methods.* Washington: Resources for the Future.

Pagiola, S., G. Acharya, and J.A. Dixon, Forthcoming. *Economic Analysis of Environmental Impacts.* London: Earthscan.

Specific techniques

Mitchell, R.C., and R. Carson. 1989. *Using Surveys to Value Public Goods: The Contingent Valuation Method.* Washington: Resources for the Future.

Hanley, N., R.E. Wright, and V. Adamowicz. 1998. "Using choice experiments to value the environment." *Environmental and Resource Economics*, **11**(3-4), pp.413-428.

Valuation techniques: Applications

McCracken, J.R., and H. Abaza. 2001. *Environmental Valuation: A Worldwide Compendium of Case Studies.* London: Earthscan.

EVRI. 2004. *Environment Valuation Reference Inventory.* Environment Canada. Available at www.evri.ca.

Forests

Bishop, J.T. 1999. *Valuing Forests: A Review of Methods and Applications in Developing Countries.* London: IIED.

Merlo, M., and L. Croitoru (eds.) . Forthcoming. *Valuing Mediterranean Forests: Towards Total Economic Value.* Wallingford: CABI Publishing.

Wetlands

Barbier, E.B., M. Acreman, and D. Knowler. 1997. *Economic Valuation of Wetlands.* Cambridge: IUCN.

Watersheds

Aylward, B. 2004. "Land use, hydrological function and economic valuation." In M. Bonnell and L.A. Bruijnzeel (eds.), *Forests, Water and People in the Humid Tropics.* Cambridge: Cambridge University Press.

Kaiser, B., and J. Roumasset. 2002. "Valuing indirect ecosystem services: The case of tropical watersheds." *Environment and Development Economics*, **7**, pp.701-714.

Coral reefs

Cesar, H.S.J. (ed.). 2000. *Collected Essays on the Economics of Coral Reefs.* Kalmar: CORDIO.

Water

Young, R.A., and R.H. Haveman., 1985. "Economics of water resources: A survey." In A.V. Kneese and J.L. Sweeney (eds.), *Handbook of Natural Resource and Energy Economics.* Vol.II. Amsterdam: North Holland.

Carbon storage

Fankhauser, S. 1995. *Valuing Climate Change: The Economics of the Greenhouse.* London: Earthscan.

Non-timber forest products

Bishop, J.T. 1998. "The economics of non timber forest benefits: An overview." Environmental Economics Programme Paper No.GK 98-01. London: IIED.

Lampietti, J., and J.A. Dixon. 1995. "To see the forest for the trees: A guide to non-timber forest benefits." Environment Department Paper No.13. Washington: World Bank.

Recreation

Herriges, J.A., and C.L. Kling (eds.). 1999. *Valuing Recreation and the Environment: Revealed Preference Methods in Theory and Practice.* Northampton: Edward Elgar.

Biodiversity for medicinal or industrial uses

Barbier, E.B., and B.A Aylward. 1996. "Capturing the pharmaceutical value of biodiversity in a developing country." *Environmental and Resource Economics*, **8**(2), pp.157-191.

Cultural benefits

Navrud, S., and R.C. Ready (eds.). 2002. *Valuing Cultural Heritage: Applying Environmental Valuation Techniques to Historic Buildings, Monuments and Artifacts.* Cheltenham: Edward Elgar.

Market-based approaches to conservation

Pagiola, S., N. Landell-Mills, and J. Bishop (eds.). 2002. *Selling Forest Environmental Services: Market-based*

Mechanisms for Conservation and Development. London: Earthscan.

Sources for specific examples used in this paper

Figure 3 and Box 4:

Croitoru, L., and M. Merlo. (forthcoming). "Mediterranean forest values." In M. Merlo and L. Croitoru (eds.), *Valuing Mediterranean Forests: Towards Total Economic Value.* Wallingford: CABI Publishing.

Figure 6:

World Bank, 1996. "Croatia Coastal Forest Reconstruction and Protection Project: Staff Appraisal Report." Report No.15518-HR. Washington: World Bank.

Figure 8:

Carret, J.-C., and D. Loyer. 2003. "Comment financer durablement le réseau d'aires protégées terrestres á Madagascar ? Apport de l'analyse économique." Paris: AFD and World Bank.

Box 3:

Bockstael, N.E., A.M. Freeman, III, R.J. Kopp, P.R. Portney, and V.K. Smith. 2000. "On Measuring Economic Values for Nature." *Environmental Science & Technology,* **34**, pp.1384-1389.

Constanza, R., R. d'Arge, R. de Groot, S. Farber, M. Grasso, B. Hannon, K. Limburg, S. Naeem, R.V.

O'Neill, J. Paruelo, R. G. Raskin, P. Sutoon and M. van den Belt. 1997. "The Value of the World's Ecosystem Services and Natural Capital." *Nature,* **387**, pp.253-260.

Toman, M. 1998. "Why not to calculate the value of the world's ecosystem services and natural capital." *Ecological Economics,* **25**, pp.57-60.

Box 5:

Pagiola, S., and G. Platais. Forthcoming. *Payments for Environmental Services: From Theory to Practice.* Washington: World Bank.

Table 1:

Millennium Ecosystem Assessment.

Table 2:

Pagiola, S., and G. Platais. Forthcoming. *Payments for Environmental Services: From Theory to Practice.* Washington: World Bank.

Adjusted Net Savings estimates cited in Chapter 4:

Hamilton, K., and M. Clemens. 1999. "Genuine savings rates in developing countries." *World Bank Economic Review,* **13**(2), pp.333-356.

World Bank. 2004. *World Development Indicators 2004.* Washington: World Bank.